John Chamberlayne

The Natural history of coffee, thee, chocolate, tobacco

In four several sections

John Chamberlayne

The Natural history of coffee, thee, chocolate, tobacco
In four several sections

ISBN/EAN: 9783337257286

Printed in Europe, USA, Canada, Australia, Japan

Cover: Foto ©berggeist007 / pixelio.de

More available books at **www.hansebooks.com**

THE
Natural Hiſtory

OF

C O F F E E,   C H O C O L A T E,
T H E E,   T O B A C C O.

In four ſeveral Sections ;

WITH A

T R A C T

OF

E L D E R and J U N I P E R - B E R R I E S,

Shewing how Uſeful they may be in Our

C O F F E E - H O U S E S :

And alſo the way of making

M U M,

With ſome Remarks upon that L I Q U O R.

Collected from the Writings of the beſt Phyſicians,
and Modern Travellers.

L O N D O N :

Printed for *Chriſtopher Wilkinſon,* at the *Black Boy* over
againſt St. *Dunſtan's* Church in *Fleetſtreet.* 168.

THE
Natural Hiſtory
OF
COFFEE.

SECT. I.

COFFEE is ſaid to be a ſort of *Arabian* Bean, called *Bon*, or *Ban* in the Eaſtern Countries, the Drink made of it is named *Coava*, or *Chaube* over all the *Turkiſh* Dominions. *Proſper Alpinus* (who liv'd Alpinus de Plant. Ægy tiac. p. 26. ſeveral years in *Ægypt*) aſſures us that he ſaw the Tree it ſelf, which he compares to our *Spindle Tree*, or *Prickwood*, only the Leaves were a little thicker, and harder, beſides continually Green. This Tree is found in the Deſarts of *Arabia*, in ſome parts of *Perſia* and *India*, the Seed or Berry of which is called by the Inhabitants *Buncho*, *Bon*, and *Ban*, which being dry'd, and boyl'd with Water, is the moſt Univerſal Drink in all the *Turkiſh*, and ſeveral Eaſtern Countries, where Wine is publickly forbid; it has been the moſt antient Drink of

Perſia) is very jocoſe and merry, when he comes to deſcribe the famous Coffee-Houſe of that City; he ſays, that the wiſe *Sha Abas* obſerving great numbers of *Perſians* to reſort to that Houſe daily, and to quarrel very much about State-affairs, appointed a *Moullah* to be there every day betimes to entertain the *Tobacco-whiffers*, and *Coffee-quaffers* with a point of Law, Hiſtory, or Poetry; after which, the *Moullah* riſes up, and makes Proclamation that every man muſt retire, and to his buſineſs: upon which they all obſerve the *Moullah*, who is always liberally entertain'd by the Company. *Olearius*

Olearius, *Ambaſſadors Travels of* Perſia. lib. 6. p. 224. does alſo ſpeak of the great diverſions made in their *Coffee-Houſes* of *Perſia* by their Poets, and Hiſtorians, who are ſeated in a high Chair, from whence they make Speeches, and tell Satyrical Stories, playing in the mean time with a little ſtick, and the ſame geſtures, as our Juglers and Legerdemain-men do in *England*.

As for the qualities and nature of *Coffee*, our own Countryman, Dr. *Willis*, has publiſh'd a very rational Account, whoſe great Reputation and Authority are of no ſmall force; he ſays, that in ſeveral Headachs, Dizzineſs, Lethargies, and Catarrhs, where there is a groſs habit of Body, and a cold heavy Conſtitution, there *Coffee* may be proper, and ſuccefsful; and in theſe caſes he ſent his Patients to the *Coffee-Houſe* rather than to the Apothecaries Shop: but where the temperament is hot, and lean, and active, there *Coffee* may not be very agreeable, becauſe it may diſpoſe the Body to inquietudes, and leanneſs. The Dr. makes one unlucky obſervation of this Drink, which I am afraid will cow our Citizens

Dr. Willis *Pharmaceut. Rat.* p. 1.

from

from ever medling with it hereafter, that it often makes
men Paralytick, and does so slacken their strings, as
they become unfit for the sports, and exercises of the
Bed, and their Wives recreations; to confirm which, I
will quote here two Precedents out of the most Learned
Olearius, who says; that the *Persians* are of an opinion
that *Coffee* allays their natural heat, for which reason
they drink it, that they may avoid the charge, and inconveniences of many Children : nay, the *Persians* are
so far from dissembling the fear they have thereof. that
some of them have come to the *Holstein* Physician of
that Embassy, for Remedies to prevent the multiplica-
tion of Children, but the Doctor being a merry bold
German, answered the *Persians,* that he had rather help
them to get Children, than to prevent them.   This most
famous *Olearius* (that made so many curious, and ac-
curate Observations in his Travels) tells us of a *Persian*
King, named *Sultan Mahomet Caswin,* who Reigned in
Persia before *Tamerlane's* time, that was so accustomed
to drinking of *Cahwa,* or *Coffee,* that he had an uncon-
ceivable aversion to Women , and that the Queen stan-
ding one day at her Chamber Window, and perceiving
they were about gelding a Horse, ask'd some standers
by, why they treated so handsom a Creature in that
manner ; whereupon answer was made her, that he was
too fiery and mettlesome, therefore they resolv'd to de-
prive him of his generative faculty : the Queen reply'd,
that trouble might have been spar'd, since *Cahwa,* or
Coffee, would have wrought the same effect, the experiment being already try'd upon the King her Husband.
This King left a Son, call'd *Mahomet,* after him, as our
most grave and faithful Travelker does assure us, who
being come to the Crown, commanded that great Poet
Hakim Fardausi, to present him with some Verses, for
every one of which the *Sophy* promised him a Ducat ;

O'earius, *Am-
bassadors Tra-
vels through
Persia. lib 5.*

Olearius. *t
b Celas T
s tho g
na. b
p. 24.*

the Poet in a ſhort time made ſixty thouſand, which at this day are accounted the beſt that ever were made in *Perſia*, and *Hakim Fardauſi* eſteem'd the Poet *Laureat* of the Eaſt ; the Treaſurers thinking it too great a ſum for a Poet, would have put him off with half, whereupon *Fardauſi* made other Verſes, wherein he reproach'd the King with Avarice, and told him, he could not be of Royal Extraction, but muſt be rather deſcended from a Shoemaker, or a Baker: *Mahomet* being netled, made complaint to the Queen his Mother, who ſuſpecting that the Poet had diſcovered her Amours, ingeniouſly confeſſed to the King her Son, that his Father being Impotent through his exceſſive drinking of *Cahwa*, or *Coffee*, ſhe fancied a Baker belonging to the Court, and ſaid, if it had not been for the Baker, the young King had never been what he was; ſo left the buſineſs ſhould take wind, the Poet got his full reward. But let us return a little into our old ſerious road.

Coffee is ſaid to be very good for thoſe, that have taken too much Drink, Meat, or Fruit, as the Learned *Schroder* will inform you, as alſo againſt ſhortneſs of Breath, and Rheum, and it is very famous in old obſtructions, ſo that all the *Ægyptian*, and *Arabian* Women, are obſerv'd to promote their Monthly courſes with *Coffee*, and to tipple conſtantly of it all the time they are flowing, for which we have the undoubted authority of *Proſper Alpinus*, who ſpent ſeveral years amongſt them. It is found to eaſe the running Scorbutick Gout, or Rheumatiſm, as *Mollenbroccius* has affirm'd.

As for the manner of preparing *Coffee*, it is ſo eaſie, and ſo commonly known, that we need not mention it, only we may obſerve, that ſome of the *Aſiatick* Nations make their *Coffee* of the Coat, or Husk of the Berry, which they look upon to be much ſtronger, and more efficacious than the Berry it ſelf, ſo that they take

Schroder's *Append.* p. 24.

Proſp. Alpinus *de Med. Ægytor.* l. 4. *de Plant. Ægyptiac.* ap. 118. *ad* p. 122.

Mollenbrock. *de Arthrit. baga ſcorbul.* p. 114.

a lefs quantity of it; but the *Europæans* do peel and take off the outward skin of the Berries, which being fo prepar'd, are Bak'd, and Burnt, afterwards grinded to Powder ; one Ounce of which they mix commonly with a Pint and a half of hot Water, which has been boyl'd half away, then they are digefted together, till they are well united.

The *Laplanders* prepare a very good Drink out of *Juniper-Berries*, which fome prefer before either *Coffee*, or *Thee*, of which Berries we will Difcourfe in a Tract at the end of thefe Sheets.

Hiftory of Lap- land.

THE

Natural Hiſtory

OF

THEE.

SECT. II.

THIS Herb *Thee* is commonly found in *China*, *Japan*, and ſome other *Indian* Countries, the *Chineſes* call it *Thee*, the *Japonians Tchia*, that of *Japan* is eſteem'd much the beſt, one pound

Nicol. Tulpii
obſervat. Med.
lib. 4. c. 60.

of it being commonly ſold for 100 pounds, as *Tulpius* informs us from ſeveral great men, that have been Ambaſſadors, and Reſidents in thoſe parts, ſo that moſt of the *Thee*, which is brought into *Europe*, comes from *China*, and that too of the worſt kind, which cannot but decay in ſo long a Voyage, for the Dutch have been obſerv'd to dry a great quantity of *Sage*, whoſe Leaves being rowl'd up like *Thee*, were carried into *China* by

Oldenburgs
Philoſ. Tranſ-
act. n. 14.

them under the name of a moſt rare *Europæan* Herb, for one pound of this dry'd *Sage* the *Dutch* receiv'd three pounds of *Thee* from the *Chineſes*, as *Thevenot* informs

us,

us, there is a great Controverſie amongſt the Herbaliſts, to what Claſſis this *Thee* may be reduc'd, *Bontius* compares it to the Leaves of our Wild *Daiſy*; for which *Simon Pauli* is very angry with him, and gives very ſtrong Arguments, that *Thee* is the Leaves of a ſort of *Myrtle*, for out of the Leaves of *Myrtle*, a Liquor may be made, reſembling *Thee* in all qualities, therefore the Jeſuite *Trigautius* is of an opinion, that ſeveral of our *European* Forreſts and Woods do abound with a true *Thee*, it being obſerv'd to grow in great plenty in *Tartary* (which lies under the ſame Climate with many Countries of *Europe*,) from whence, ſome Learned men think, it came Originally, for it has not been long known to the *Chineſes*, they having no antient name, or Hieroglyphick Characters for *Thee*, and *Cha* being an antient *Tartarian* Word, beſides it is known to ſeveral Merchants, that a great quantity of *Thee* is brought yearly out of *Tartary* into *Perſia*, we are all acquainted with the ſeveral great Conqueſts, which the *Tartars* have made in *China*, ſo that the *Chineſes* have had ſeveral opportunities of learning the uſe of *Thee* from the *Tartars*, in whoſe Country it is obſerv'd to be in great plenty, and of little value; yet the Inhabitants of *China* and *Japan* have a great eſteem, and opinion of it, where they are as much employ'd, and concern'd for their Harveſt of *Thee*, (which is in Spring) as the *Europeans* are for their Vintage, as ſeveral Jeſuits inform us in their Obſervations of *China*: for the Noblemen, and Princes of *China* and *Japan*, drink *Thee* at all hours of the Day, and in their Viſits it is their whole Entertainment, the greateſt Perſons of Quality Boyling, and Preparing the *Thee* themſelves, every Palace, and Houſe, being furniſht with convenient Rooms, Furnaces, Veſſels, Pots and Spoons for that purpoſe, which they value at a higher rate than we do Diamonds, Gems, and Pearls, as *Tulpius*

B   aſſures

Bontius de Medicina Indor. lib. 2. p. 97.

Simon Pauli de Thee, p. 19, 20.

Trigautius de regno Chinæ. lib. 3.

Simon Pauli de Thee. P. 25.

Olearius, Ambaſſadors Travels in Perſia. p. 241.

Philoſ. Transact. N. 49.

Nicol Tulpii Obſervat. Med. lib. 4 c. 60.

assures us from the relations of several great *Dutch-men*, who travell'd *China* in the Quality of Ambassadors, and made great Observations of those rich Stones, and Woods,out of which the aforesaid Materials were made.

As for the Qualities and Vertues of *Thee*, these few following Observations may give satisfaction, that it makes us active and lively, and drives off sleep, every Drinker of it cannot but be sensible. The great Jesuit *Alexander de Rhodes*, always Cur'd himself of a Periodical pain of his Head by *Thee*, and having often occasion to sit up whole Nights in *China* to take the Confessions of dying People,he found the great benefit of *Thee* in those great watchings,so that he was always as vigorous, and fresh the next day, as though he had rested all night ; nay, he says, that he sate up six nights together by the assistance of *Thee*. *Kircher* himself took notice of *Thee* for clearing the Head, and opening the Urinary passages ; and it was observ'd by those concern'd in the *Dutch* Embassy to *China*, that the *Chinefes* did spit very little, and were seldom subject to the Stone, and Gout, which their Physicians imputed to their frequent Drinking of *Thee*: it is a common Proverb in *Japan, Illenè sanus non fit ? Bibit de optimâ Tsjâ*, What, is not he well ? He Drinks of the best *Thee*. I know some that Celebrate good *Thee* for preventing Drunkenness, taking it before they go to the Tavern, and use it also very much after a Debauch, *Thee* being found so friendly to their Stomachs, and Heads: several Ambassadors find the advantage of it in preserving them from the accidents and inconveniences of a bad Foreign Air ; but that which gives the greatest commendation to *Thee*, is the good Character which our famous Country-man, Mr. *Boyl*, gives of it in his Experimental Philosophy, where he says, that it deserves those great praises which are commonly bestow'd upon it. Yet *Simon Pauli* exclaims against the use

Alexander de Rhodes Voyages & missions Apostoliques.

Kircheri China illustrata. lib. 4.

Thevenotts Histor. legat. Batavor. in China. Tom. 3. Philosoph.Transact. N. 14. Varenius descript. Regni Japon. c. 23. p. 151.

Boyles exper. Philos. p. 94.

Simon Pauli de Thee. p. 67.

ufe of *Thee*, as a great dryer, and promoter of old Age, and as a thing unnatural, and foreign to the *European* Complexions. But *Schroder* anfwers *Pauli* very mild- ly, fuppofing him to fpeak only of the abufe, and extra- vagant management of *Thee*; for otherwife *Rheubarb*, *China*, *Saffafras*, and *Saunders*, fhould be banifht from our Shops by the fame reafon, they being Dryers, and foreign to us *Englifh-men*, therefore we may conclude *Thee* innocent, and beneficial. *Schroderi Ap- pend. ad Phar- macop.* p 28.

The *Chinefes* gather the Leaf in the Spring one by one, and immediately put them to warm in an Iron Kettle o- ver the Fire, then laying them on a fine light Mat, rolls them together with their Hands; the Leaves thus roll'd are again hang'd over the Fire, and then roll'd clofer to- gether, till they are dry, then put up carefully in Tin Veffels to preferve them from moifture : thus they pre- pare the beft Leaves, that yield the greateft rates, but the common ordinary ones are only dry'd in the Sun, yet in the Shade is doubtlefs much better, (as the inge- nious Author of *Vinetum Britannicum* does well obferve) the Sun having a great power to attract the vertue out of any Vegetable after its feparation from its nourifher, the Earth. One fpoonful of this prepar'd *Thee* is enough for one quart of boyl'd Water. *Vinet. Britan.* p. 140.

There are feveral ways and methods for preparing *Thee*. The *Japonians* powder the Plant upon a Stone, and fo put it into hot Water. The *Chinefes* boyl the Leaves with Water and a little Sugar. Some *Europeans* make Tinctures, Infufions, Conferves, and Extracts of *Thee*. The *Tartars* are obferv'd to boyl their *Thee* in Milk with a little Salt, which way they think is the ve- ry beft. *Nicol. Tulpii obfervat. Med. lib.* 4. c. 50.

Thevenots Hi- ftor. legat. Bel- gic. ad finerfi- um regem.

The Inhabitants of *Carolina* prepare a Liquor out of the Leaves of an *American* Tree, which is very like *Thee*, and equal to it in every refpect, Dr. *Mundy* obferves that *Dr. Mundy de potuantis.* p. 3;3.

B 2   the

the Inhabitants of *Florida* have an old cuftom, before they go into the Field to War, of Drinking a Liquor in a great publick Affembly, which he that Vomits up, is judg'd unfit for that Warlike Expedition, and is condemn'd to ftay at home in difgrace; but when he has learnt to carry off the Liquor, then he is admitted to be a lawful Soldier: Now *Thee* it felf when given in a large dofe, and in a ftrong Decoction, does often prove Vomitive, as I my felf have obferv'd feveral times.

Some make Decoctions of the Roots of *Avens, Galanga, Coriander*, and *Anifeeds, Sarfa, China, Saunders*, of the Leaves of *Sage, Betony, Rorifmary*, &c, which they do extol above *Thee* or *Coffee*.

THE

THE
Natural Hiſtory
OF
CHOCOLATE.

SECT. III.

HAving given a ſhort Natural Hiſtory of two things, which are ſo univerſally us'd in the Eaſtern part of the World, we now come to treat briefly of two more, which are generally us'd in the Weſtern: Firſt of *Chocolate,* of which the *Cacao,*or *Cacaw-nut,* being the principal Ingredient, a ſhort Account of it cannot be improper,this *Nut,* or rather the Seed, or Kernel of the *Nut,* as Mr. *Hughes* obſerves, is of the bigneſs of a great *Almond;* in ſome of theſe Fruits there are a dozen, in ſome 20, in others 30, or more of theſe Kernels, or *Caco's,* which are well deſcrib'd by the Ingenious and Learned Dr. *Grew,* when theſe Kernels are cured they become blackiſh, and are compar'd to a Bullocks Kidney, cut into Partitions;there is great variety in them, by reaſon of the difference of Soyls and Climates where they grow: the Tree is ſaid to be as large as our *Engliſh Plumb-trees.* the Leaves ſharp-pointed,compar'd by ſome Travellers to the Leaves

Hughes Ameri-can *Phyſician,* p. 115.

Dr.Grew *Muſ. Reg. Soc. Angl.* p 204.

of

Piſo *in Hiſtor.*
Nat. Indiæ
ntrinſqut.

Hughes Ame-
rican *Phyſici-*
an, p. 112.

Joſ. Acoſta *In-*
dor. Hiſtor.
lib. 4. c. 22.

Gages Survey
of the *weſt In-*
dies. Chap. of
Chocolate.

of *Cheſnut*; by the Curious *Piſo* to the Leaves of an *Orange*, the Flower of a *Saffron* colour, upon the appearance of which, the Fruit appears upon the Branches as Apples: This Tree grows in ſeveral parts of *America,* as in *Nicaragua, New Spain, Mexico, Cuba,* and in *Jamaica,* eſpecially at Collonel *Larrington*'s Quarters, or Plantations; they proſper beſt in low, moiſt, and fat ground, and are as ſquarely, and orderly ſet, as the *Cherry* Trees in *Kent,* or *Worceſterſhire:* they commonly bear within 7 years, and then twice every year, the firſt Crop between *January* and *February,* the other between *May* and *June.* The Inhabitants have ſo great a value for them, as that they ſecure them with the ſhades of *Plantane* and *Bonona* Trees, againſt the injuries of their fiery Sun, and do uſe the Kernels inſtead of Money, both in their Traffick, and Rewards; as the great Jeſuit, *Joſephus Acoſta,* obſerv'd, when he was ſent into *America:* The *Indians* look upon their *Chocolate* as the greateſt delicacy for extraordinary Entertainments. *Montezuma* is ſaid to have Treated *Cortez* and his Soldiers with it; and you can ſcarce read an *American* Traveller, but he will often tell you of the magnificent Collations of *Chocolate,* that the *Indians* offer'd him in his Paſſage and Journies through their Country: as Mr. *Gage* (who Travell'd many years in *America*) informs us, the *Spaniards* do conſtantly drink *Chocolate* in their Churches, at *Mexico* and *Chiapa,* of which they being once forbid, did Mutiny, and commit great Outrages, till their Cuſtom was reſtor'd them. The *Indians,* and *Chriſtians,* in the *American* Plantations, have been obſerv'd to live ſeveral Months upon *Cacao Nuts* alone, made into a Paſte with Sugar, and ſo diſſolv'd in Water; I my ſelf have eaten great quantities of theſe Kernels raw, without the leaſt inconvenience: and have heard, that Mr. *Boyle,* and Dr. *Stubbs,* have let down into their Stomachs ſome

pounds of them raw without any moleftation; the Sto-
mach feems rather to be fatiated, than cloy'd with them,
which is an Argument they are foon diffolv'd, and digefted. The *Spaniards* do not fcruple to eat them upon
their great Faft days.

The *Indians* at firft made their *Chocolate* of the *Nut* a-
lone without any addition, unlefs fometimes *Pepper*,
and *Maiz*, or *Indian* Wheat, and in *Jamaica* at this day,
as Mr. *Hughes* obferves, there is a fort of *Chocolate*, *Hughes Ameri-*
made up only of the Pafte of the *Cacao* it felf, and this *can Phyfician.*
he efteems to be one of the beft forts of *Chocolate*. Dr. *Dr. Stubbs In-*
Stubbs, who was a great Mafter of the *Chocolate* Art, *dian Nectar.*
did not approve of many Ingredients, befides the *Cacao*
Nut; that *Chocolate* which the Doctor prepared for His
Majefty, had double the quantity of the *Cacao* Kernel to
the other Ingredients: In the common fort the *Cacao*
Nuts may take up half the Compofition, according to
Pifo, in the worft a third part only. As to the other *Pifo Nat. Hi-*
Ingredients for making up *Chocolate*, they may be vari- *ftor. Indor.*
ed according to the conftitutions of thofe that are to
drink it; in cold conftitutions *Jamaica Pepper*, *Cinna-*
mon, *Nutmegs*, *Cloves*, &c. may be mixt with the *Cacao*
Nut: fome add *Musk*, *Ambergreafe*, *Citron*, *Lemmon-*
Peels, and Odoriferous Aromatick Oyls: In hot Con-
fumptive tempers you may mix *Almonds*, *Piftacho's*, &c.
fometimes *China*, *Sarfa*, and *Saunders*; and fometimes
Steel and *Rheubarb* may be added for young green Ladies. Mr. *Hughes* gives us very good advice, in telling *Hughes Ame-*
us, that we may buy the beft *Chocolate* of Seamen and *rican Phyfici-*
Merchants, who bring it over ready made from the *Weft* *an. p. 111.*
Indies; his reafon is none of the worft, which is this, let
the *Cacao* Kernels be never fo well cur'd in the *Weft In-*
dies, and ftowed never fo carefully in the Ship, yet by
their long tranfportation, and by the various Airs or
Climates they are often fpoil'd, their natural Oylinefs
tending

tending much to putrefaction: from whence I have heard several complain in *England*, that their *Chocolate* made up here does often prove musty, and will settle much to the bottom of the Dish, which is a certain sign, says the Learned Dr. *Stubbs*, that the *Nuts* are either faulty, or not well beaten, and made up. The best *Cacao Nuts* are said to come from *Carraca* or *Nicaragua*, out of which Dr. *Stubbs* prepar'd *Cholocate* for the King; yet the Doctor commends the *Cacao Nuts* of *Jamaica*, which were first Planted there by the *Spaniards*. That you may know how to Prepare your *Chocolate*, I will give you a short direction, if you intend to make it up your self; consult your own constitution and circumstances, and vary the Ingredients according to the Premises, for I cannot give a Receipt to make up the mass of *Chocolate*, which will be agreeable, and proper to all Complexions; yet in the Composition of it, you must remember to appoint the *Cacaw Kernel* for the fundamental and principal Ingredient: as for the managing the *Cacao Nut*, Dr. *Stubbs*, and Mr. *Hughes*, have publisht most excellent instructions, how you must peel, dry, beat and scarce it very carefully, before you beat it up into a mass with other simples: as for the great quantity of Sugar which is commonly put in, it may destroy the Native and Genuine temper of the *Chocolate*, Sugar being such a corrosive salt, and such a Hypocritical Enemy to the Body. *Simon Pauli* (a Learned *Dane*) thinks Sugar to be one cause of our *English* Consumptions; and Dr. *Willis* blames it as one cause of our Universal Scurvy's: therefore when *Chocolate* produces any ill effects, they may be often imputed to the great superfluity of its Sugar, which often fills up half its Composition. For preparing the Drink of *Chocolate*, you may observe the following measures. Take of the mass of *Chocolate*, cut into small pieces, one ounce, of Milk and Water well boyl'd

Dr. Stubbs In-dian Nectar.

Dr. Stubbs In-dian Nectar. and Mr. Hughes American Physician.

Simon Pauli quadripart Botan.
Dr. Willis de Scorbuto.

boyl'd together, of each half a pint, one yolk of an Egg well beaten, mix them together, let them boyl but gently, till all is diſſolved, ſtirring them often together with your Mollinet, or *Chocolet* Mill; afterwards pour it into your Diſhes, and into every Diſh put one ſpoonful of Sack.

As for the vertues and effects of the *Cacao Nut*, or *Chocolate*, all the *American* Travellers have written ſuch Panegyricks, and ſo many Experimental Obſervations, that I ſhould but degrade this Royal Liquor, if I ſhould offer at any; yet I think two or three Remarks upon it cannot be unſuitable to this little Hiſtory: ſeveral of theſe curious Travellers, and Phyſicians, do agree in this, that the *Cacao Nut* has a wonderful faculty of quenching thirſt, allaying Hectick heats, of nouriſhing and fatning the Body. Mr. *Gage* acquaints us, that he drank *Chocolate* in the *Indies* two or three times every day for twelve years together, and he ſcarce knew what any Diſeaſe was in all that time, he growing very fat: ſome object it is too oily and groſs, but then the bitterneſs of the *Nut* makes amends, carrying the other off by ſtrengthening of the Bowels. Mr. *Hughes* informs us, that he liv'd at Sea for ſome Months of nothing but *Chocolate*, yet neither his ſtrength, nor fleſh were diminiſhed: he ſays our *Engliſh* Seamen are very greedy of it when they come into any *Indian* Port, and ſoon get plump countenances by the uſe of it. Mr. *Hughes* himſelf grew very fat in *Jamaica* by the vertue of the *Cacao Nut*; ſo he judges it moſt proper for Lean, Weak, and Conſumptive Complexions: it may be proper for ſome breeding Women, and thoſe perſons that are Hypocondriacal, and Mellancholly. The induſtrious Dr. *Mundy* gives a notable example of the effect of *Chocolate*, he ſays, that he knew a Man in a deſperate Conſumption, who took a great fancy for *Chocolate*, and his Wife out of complaiſance drank it often with him: the conſequence was

Joh. de Laet. Hiſtor. Indor. Piſo ʳ. Hiſtor. Indor. Herbar. Mexican. Benzonus Hiſtor. Indor. Occident. &c.

Gages Survey of the Weſt Indies. Chap. of Chocolate.

Hughes American Phyſician, p. 147.

Dr. Mundy de potulentis. p. 350.

C   this,

this, the Husband recover'd his health, and his Wife
brought afterwards to Bed of three Sons at one Birth.

   The great use of *Chocolate* in Venery, and for supply-
ing the Testicles with a Balsam, or a Sap, is so ingeni-
ously made out by one of our Learned Countrymen already, that I dare not presume to add any thing after so
accomplisht a Pen ; though I am of an opinion, that I
might treat of the Subject without any immodesty, or
offence. *Gerson* the Grave *Roman* Casuist, has writ *de
Pollutione Nocturnâ*, and some have defended Fornication
in the Popish Nunneries; Hysterical fits, Hypocondriacal Melancholy, Love Passions, Consumptive Pinings
away, and Spermatical Feavers, being instances of the
necessity hereof, natural instinct pointing out the Cure:
We cannot but admire the great prudence of *Moses*, who
severely Prohibited, that there should be no Whore a-
mongst the Daughters of *Israel*, yet that most wise Le-
gislator took great care for their timely Marriage : upon
these very accounts the Casuists defend the Protestant
Clergy in their Marriages. And *Adam* is commanded
in Paradise to Encrease and Multiply, therefore I hope
this little excursion is pardonable, being so adæquate to
this Treatise of *Chocolate*; which if *Rachel* had known,
she would not have purchas'd *Mandrakes* for *Jacob*. If
the Amorous and Martial *Turk* should ever taste it, he
would despise his Opium. If the *Grecians* and *Arabians*
had ever try'd it, they would have thrown away their
Wake-Robins, and their *Cuckow-Pintles*; and I do not
doubt, but you *London* Gentlemen, do value it above all
your *Cullises* and *Jellies*, your *Anchoves*, *Bononia Sausages*,
your *Cock*, or *Lamb-stones*, your *Soys*, your *Ketchups* and
Caveares, your *Cantharides*, and your *Whites* of *Eggs*,
are not to be compared to our rude *Indian*; therefore you
must be very courteous and favourable to this little
Pamphlet, who tells you most faithful Observations.
                                                    The

The induſtrious Author of the *Vinetum Britannicum* Vinet. Britan. makes a Quære, whether the Kernel of the *Wallnut* may P. 139. not ſupply the defect of the *Cacao*, if well Ground. Dr. *Grew* thinks, that for thoſe that drink *Chocolate* at Dr. Grew's *Coffee-Houſes* without any Medicinal reſpect, there is no Muſ. Reg. Soc. doubt, but that of *Almonds* finely beaten, and mixed P. 205. with a due proportion of Spices, and Sugar, may be made as pleaſant a Drink as the beſt *Chocolate.*

C 2   THE

THE

Natural Hiſtory

OF

TOBACCO.

SECT. IV.

TOBACCO is reckon'd by the beſt Herbaliſts to be a Species, or ſort of *Henbane*, proper to the *American* Regions, as *Dodonæus* and *Simon Pauli*; yet ſome *Botaniſts* will have it a Native of *Europe*, and reduce it to ſeveral of our Claſſes: but I will not trouble you with this Controverſie, only we may take notice, that *Thevet* did firſt bring the ſeed of *Tobacco* into *France*, though *Nicot* the *French* Ambaſſador in *Portugal* (from whom it is call'd *Nicotiana*) was the firſt that ſent the Plant it ſelf into his own Country. *Hernandes de Toledo* (who Travell'd *America* by the Command of *Philip* II.) having ſupply'd *Spain* and *Portugal* with it before. Sir *Francis Drake* got the Seed in *Virginia*, and was the firſt that brought it into *England*; yet ſome give Sir *Walter Rawleigh* the honour of it, ſince which time it has thriven very well

Dodonæus Herbal. Simon Pauli *quadripart. Botan. & lib. de Tabaco.*

Hernandez *Hiſtor. American.* Purchas Voyages into *America.*

in

in our *English* Soil: a great quantity of it grows yearly in several Gardens about *Westminster*, and in other parts of *Middlesex*.   It is planted in great plenty in *Gloucester*, *Devonshire*, and some other Western Countries; his Majesty sending every year a Troop of Horse to destroy it, lest the Trade of our *American* Plantations should be incommoded thereby: yet many of the *London* Apothecaries make use of *English Tobacco* in their Shops, notwithstanding the vulgar Opinion that this Herb is a Native of *America*, and foreign to *Europe*: yet *Libavius* assures us, that it grows naturally in the famous *Hercynian* Forrest of *Germany*.   If this was true, we would no longer call it *Tobacco* from the Island of *Tobago*.   The names of it are so various, as they would glut the most hungry Reader.   The *Americans* style it *Picielt*; in *Nova Francia*, *Petum*; in *Hispaniola*, *Cozolba*; in *Virginia*, *Uppuyoc*; at *Rome*, *Herba Sancta Crucis*; in some parts of *Italy*, *Herba Medicea*; in *France*, *Herba Reginæ*, as you may read in *Magnenus* and *Neander*: but let it be of *Magnenus de* what name or kind it will, I am confident, that it is of *Tabaco.Nander* the poysonous sort, for it Intoxicates, Inflames, Vomits, *Tabacolog.* and Purges; which Operations are common to poysonous Plants, as to *Poppeys*, *Nightshades*, *Hemlocks*, *Monks. hood*, *Spurges*, and *Hellebores*, that will produce the like effects: besides, every one knows that the Oyl of *Tobacco* is one of the greatest Poysons in nature, a few drops of it falling upon the tongue of a Cat, will immediately throw her into Convulsions, under which she will die. This Dr. *Willis* assures us to be true; the experiment *Dr. Willis* succeeded, when it was try'd before the Royal Society, as *Pharm. Rat.* the Learned Dr. *Grew* has affirmed: besides, I can speak *Dr. Grew's* it upon my own certain knowledg, having kill'd several *Muf. Reg Soc.* Animals with a few drops of this Oyl.   Yet that most *p. 352.* sagacious *Italian*, *Francisco Redi*, observes very well, *Philof. Tranf-* that the Oyl of *Tobacco* kills not all Animals, neither *act. Olden-* *burgh N. 92.*

does it difpatch thofe, it kills, in the fame fpace of time; there is a great difference between the *Tobacco* of *Brazil*, and that of St. *Chriftophers*, as to this effect: *Varino* and *Brazil Tobacco* being almoft of the fame quality and operation, whereas that of St. *Chriftophers*, *Terra Nova*, *Nieve*, St. *Martin*, have very different effects.

If we run over thofe Countries where *Tobacco* is made ufe of, we may obferve the various manners of ufing it; fome *Americans* will mix it with a Powder of Shells, to chew it, falivating all the time, which they fancy does refrefh them in their Journeys and Labours: others in *New Spain* will dawb the ends of Reeds with the Gum, or Juice of *Tobacco*, and fetting them on fire, will *Purchas* Voya-ges to *America*. fuck the fmoak to the other end. The *Virginians* were obferv'd to have Pipes of Clay before ever the *Englifh* came there, and from thofe *Barbarians* we *Europeans* have borrow'd our mode and fafhion of fmoaking. The *Moors* and *Turks* have no great kindnefs for *Tobacco*; yet when they do fmoak, their Pipes are very long, made of Reeds, or Wood, with an earthen head. The *Irifh-men* do moft commonly powder their *Tobacco*, and fnuff it up their Noftrils, which fome of our *Englifh-men* do, who often chew, and fwallow it; I know fome Perfons that do eat every day fome ounces of *Tobacco* without any fenfible alteration: frome whence we may learn, that ufe and cuftom will tame, and naturalize the moft fierce and rugged Poyfon, fo that it will become civil and *Ephem.German.* A⁰. 2. friendly to the body. We read of a *French* Ambaffador, that being in *England*, was fo indifpos'd, that he could never fleep; upon which he would often devour whole Ounces of *Opium* without being concern'd: and the *Turks* are often obferv'd to fwallow great Lumps of it, a tenth part of which would kill thofe that were not accuftomed to *Opiates*. I know a Woman in this City, that being us'd to take both the *Hellebores*, will often fwallow whole Scruples of them without the leaft motion, or operation, fo that cuftom and converfation will make the fierceft creature familiar.

As for the Culture, Harveft, Preparation, and Traffick of *Tobacco*, I will recommend you to *Neander*, where, if you are curious, you may meet with fatisfaction. I cannot omit one Story out of *Monardus*, who tells us, that the *Indian* Priefts being always confulted about the events of War, do burn the Leaves of *Tobacco*, and fucking into their mouths the fmoak by a Reed, or Pipe, do prefently fall into a Trance, or Extafie, and as foon as ever they come out of it, they difcover to the *Indians* all the fecret Negotiation, which they have had with the great *Dæmon*, always delivering fome ambiguous anfwer.

As for the qualities, nature, and ufes of *Tobacco*, they may be very confiderable in feveral cafes and circumftances; though King *James* himfelf has both Writ, and Difputed very fmartly againft it at *Oxford*, and *Simon Paulî* has Publifh'd a very Learned Book againft it. Some Anatomifts tell us moft terrible Stories of footy Brains, and black Lungs, which have been feen in the Diffections of Dead Bodies, which when Living had been accuftomed to *Tobacco*. We read that *Amurath* the Fourth did forbid the ufe of it over all the *Turkifh* Dominions, under the moft fevere Penalties; the *Turks* having an opinion amongft them, that *Tobacco* will make them Effeminate, and Barren, unfit for War and Procreation; though fome think there is a Politick defign in it, to obftruct the fale of it in the Eaftern Countries, and to prevent the *Chriftians* from eftablifhing any confiderable Traffick from fo mean a Commodity, which perhaps may be one reafon, why the Great Duke of *Mufcovy* has threatned to punifh thofe

Marginal notes:
Neander Tabacalog.
Monardus *lib.* x. *Exotic. cor. Claffi.*
Simon Paulî de abufu Tabaci. Diemerbrock. &c. Hoffman. Paulini.
Olearius, *Am. Iterf. Tabago. gh Mufcovy.*

thofe Merchants, who offer to fell any *Tobacco* in his Countries. *Scach Abas* (the Great *Sophy* of *Perfia*) Leading an Army againft the *Cham* of *Tartary*, made, Proclamation, that if any *Tobacco* was found in the Cuftody of any Soldier, he fhould be burnt alive, together with his *Tobacco*. Yet for all this it may be very beneficial to Mankind, as you will conclude from what does follow.

Dr. *Willis* recommends *Tobacco* to Soldiers, becaufe it may fupply the want of Victuals, and make them infenfible of the dangers, fatigues, and hardfhips, which do ufually attend Wars and Armies; befides. it is found to Cure Mangy, and Ulcerous Difeafes, which are frequent in Camps. I know a curious Lady in the *North*, that does very great feats in Sores and Ulcers by a Preparation of *Tabacco*. Our Learned and moft Experienc'd Countryman, Mr. *Boyle*, does highly commend *Tobacco* Clyfters in the moft violent Colick pains, which are often Epidemical in Cities, and Camps. The Renowned *Hartman* extols the Water of *Tobacco* againft Agues: And the curious Dr. *Grew* found the fuccefs of the Oyl of it in the Tooth-ach, a Lint being dip'd in it, and put into the Tooth. The effects of *Tobacco* has been very good in fome violent pains of the Head; as fome thoufands have experimented: As for the daily fmoaking of it, the ftate and circumftances of your Body muft be the beft guide, and rule; if your complexion be lean, hot, and dry, it is an argument againft it, but if cold, moift, and humoral, fubject to Catarrhs, Rheums, and Pains, then there may be a temptation to venture upon it, fo every man muft confult his own temper, and the experience of others.

Dr. Willis
Pharm. Rat.

Boyl's experiment Philofophy.

Hartman prax.
Chym
Dr. Grew Muf.
Reg. Soc. p. 252.

A

A modern *French* Author has writ a peculiar Tract of *Journal des* *Tobacco*, wherein he commends it in Convulfions, in $^{Scavans.}_{An. 16\\!1.}$ pains, and for bringing on fleep; he extols the Oyl of it in Curing Deafnefs, being injected into the Ear in a convenient vehicle, alfo againft Gouty and Scorbutical pains of the Joints, being appli'd in a liniment. *A Lixivium* of *Tobacco* often prevents the falling off of the hair, and is famous in Curing the Farcy, or Leprofie of Cattel.

D   THE

The USE of

JUNIPER

AND

ELDER-BERRIES,

IN OUR

Publick-Houfes.

THESE two *Berries* are fo Celebrated in many Countries, and fo highly recommended to the World by feveral famous Writers, and Practitioners, that they need not defire any Varnifh, or Argument from me. The fimple Decoctions of them fweetned with a little fine Sugar-Candy will afford Liquors fo pleafant to the Eye, fo grateful to the Palate, and fo beneficial to the Body, that I cannot but wonder after all thefe Charms, they have not as yet been Courted, and Ufher'd into our Publick Houfes; if they fhould once appear on the Stage, I am confident, that both the *Whig* and the *Tory*, would agree about them far better than have done about the *Medal* and *Mufhroom*: nay, the very Cynick and-Stoick himfelf, would fall in Love with

the

the Beauty, and extraordinary Vertues of thefe *Berries*, which are fo common, and cheap, that they may be pur-chas'd for little or nothing ; one Ounce of the *Berry* well cleanfed, bruis'd, and mafh'd, will be enough for almoft a Pint of Water; when they are boyl'd together, the Veffel muft be carefully ftopt : after the boyling is over, one fpoonful of Sugar Candy may be put in.

The *Juniper-tree* grows wild upon many Hills in Of the *Juniper-* *Surrey*, and *Oxfordfhire*, and upon *Juniper-Hill* near *Berry.* *Hilderfham* in *Cambridgfhire* ; befides, in feveral other Dr. Merrets parts of *England* : The *Berries* are moft commonly ga- Pinax. ther'd about *Auguft*. The *Aftrological Botanifts* advife Ray's Catalog. us to pull them, when the Sun is in *Virgo*. Plantar.

The *Juniper-Berry* is of fo great reputation in the *Nothern* Nations, that they ufe it, as we do *Coffee* and Hiftory of *Lap-* *Thee*, efpecially the *Laplanders*, who do almoft adore it. land. *Simon Pauli* (a Learned *Dane*) affures us, that thefe Simon Pauli *Berries* have perform'd wonders in the Stone, which he quadripartit. did not learn from Books, or common Fame, but from Botan. p. 536. his own obfervation and experience; for he produces two very notable examples, that being tormented with the Stone, did find incredible fuccefs in the ufe of thefe *Berries* : and if my memory does not fail me, I have heard our moft ingenious, and famous Dr. *Troutbeck*, commend a Medicine prepar'd of them in this Diftem-per. Befides *Schroder* knew a Nobleman of *Germany*, Schroder. that freed himfelf from the intolerable fymptoms of the Pharmacop. Stone by the conftant ufe of thefe *Berries* : Ask any Phyfician about them, and he will beftow upon them a much finer Character than my rude Pencil can draw. The Learned Mr. *Evelyn* will tell you what great kind- Evelyn of For-neffes he has done to his Poor fick Neighbours, with reft Trees, a Preparation of *Juniper-Berries*, who is pleas'd to p. 135. honour them with the Title of the *Forrefter's Panacæa* ; he extols them in the Wind Colick, and many other
Diftem-

Diftempers. Do but confult *Bauhinus*, and *Schroder*, the firft being the moft exact Herbal, the other the moft faithful and elaborate Difpenfatory, that ever has been publifh'd; and you will find great commendations of thefe *Berries* in Dropfies, Gravel, Coughs, Confumptions, Gout, Stoppage of the monthy Courfes, in Epilepfies, Palfies, Lethargies, in which there are often an ill appetite, bad digeftions, and obftructions.

Joh. Bauhin. Hiftor. Plantar. Schroder. Pharmacop.

Take one fpoonful of the Spirit of *Juniper-Berries*, four grains of the Salt of *Juniper*, three drops of the Oyl of *Juniper-Berries* well rectified; mix them all together, drink them Morning and Night in a Glafs of White-wine, and you will have no contemptible Medicine in all the aforementioned Difeafes.

Now it is probable, that you have both the Spirit, Salt, and Oyl of this *Berry* in a fimple Decoction of it, provided it be carefully and skilfully manag'd. If this will not fatisfie, do but read *Benjamin Scarffius*, and *Joh. Michael*, who have Publifh't in *Germany* two feveral Books of the *Juniper*, and you may meet with far more perfuafive arguments, than I can pretend to offer you.

Scarffius de Junipero. Joh. Michael Juniperet.

Of *Elder-Berries*.

The *Elder Tree* grows almoft every where, but it moft delights in Hedges, Orchards, and other fhady places, or on the moift Banks of Rivulets and Ditches, unto which 'tis thruft by the Gardeners, left by its Luxury, and importunate increafe yearly it fhould poffefs all their ground. We write here of the Domeftick, common *Elder*, not of the Mountain, the Water, or Dwarf *Elder*, ours in figure is like the *Afh*; the Leaves refemble thofe of a *Walnut Tree*, but lefs; in the top of the Branches, and Twigs, there fpring fweet and crifped umbels, fwelling with white odoriferous Flowers (in *June* before St. *Johns* Eve) which by their fall give place to a many branched *Grape*, firft green, then ruddy, laft of a black, dark, Purple Colour, fucculent and tumid with its

Dr. Blochwich Anatom. Sambuci.

winifh

winish Liquor: of all the wild Plants 'tis first covered *Joh. Bauhin.* with Leaves, and last uncloathed of them. It flourishes in *Hiſtor.Plantar.* *May, June, July,* but the *Berries* are not ripe till *Auguſt.*

As for the qualities, and vertues of *Elder-Berries,* I need say no more, but that Mr. *Ray* has given a great *Ray Catalog.* encomium of them; our Learned Dr. *Needham* com-*Plantar.* mending them in Dropsies, and some Feavers: and I have been inform'd, that the ingenious Dr. *Croon* has extoll'd a Spirit of *Elder-Berries* in an Epidemical inter-mittent Feaver. *Schroder* says, they do peculiarly re-*Schroder.* spect some Diseases, attributed to the Womb. Mr. *Eve-*Pharmacop.* *lyn* is so bountiful to his poor *Forreſter,* as to assure him, *reſt-Trees,* that if he could but learn the Medicinal Properties of the *p. 99.* *Elder Tree,* he might fetch a Remedy from every Hedg, either for Sickneſs, or Wound: the same curious Gen-tleman takes notice, how prevalent these *Berries* are in scorbutick Diſtempers, and for the prolongation of Life (so famous is the Story of *Næander.*) I have heard some praise them in Bloody Fluxes, and other Diſeaſes of the Bowels; also in several Diſtempers of the Head, as the Falling Sickneſs, Megrims, Palſies, Lethargies: they are said likewiſe to promote the monthly Inundations of Women, and to deſtroy the heat of an Eryſipelas, for which the Flowers themſelves are highly Celebrated by *Simon Pauli,* who experimented them upon himſelf with *Simon Pauli* wonderful ſucceſs. I could produce ſeveral caſes out *quadripartit.* *Botanic. p.139,* of the beſt Phyſical Writers, as *Foreſtus, Riverius, Ru-*140. *landus,* &c. where theſe *Berries* have acted their parts, even to admiration; but if you are curious, and inqui-ſitive after the qualities and nature of them, I will re-commend a Learned *German, Martyn Blochwitz,* to your *Dr. Blochwitz* reading, where you may entertain your ſelf with great *Anatom. of the* variety: Yet I have one thing ſtill to give notice of, *Elder.* that the ſame Medicine may be prepar'd out of the Spi-rit, Oyl, and Salt of this *Berry,* as you have been
taught

taught before to make out of the *Juniper-Berry*, but you may obtain them all in a ſimple Decoction, if it be well manag'd.

You have read here the great uſe of theſe two *Berries*, that are more univerſally agreeable to all tempers, palates and caſes, than perhaps any other two ſimple Medicines, which are commonly known amongſt us; ſo that ſeveral Perſons being under ill habits of Body, and upon the Frontiers of ſome lingring Diſeaſes, cannot but deſire to drink them, when they have occaſion to reſort to Publick-Houſes: yet for all this, my poor advice will certainly meet with that Fate, which does attend almoſt every thing in the World, that is, *Laudatur ab his, culpatur ab illis*; but it dreads moſt of all the *Turkey*, and *Eaſt-India* Merchant, who will condemn it in defence of their *Coffee*, and *Thee*, which have the honour of coming from the *Levant*, and *China*. Beſides, I am afraid of a laſh, or a frown from ſome young Ladies, and little Sparks, who ſcorn to eat, drink, or wear any thing, that comes not from *France*, or the *Indies*; they fancy poor *England* is not capable of bringing forth any commodity, that can be agreeable to their Grandeur, and Gallantry, as though Nature, and God Almighty, had curs'd this Iſland with the Production of ſuch things, as are every way unſuitable to the Complexions, and Neceſſities of the Inhabitants: ſo we cannot but Repartee upon theſe *A la mode* Perſons, that while they Worſhip ſo much only Foreign Creatures, they cannot but be wholly ignorant of thoſe at home. His Excellency, the moſt Acute and Ingenious Ambaſſador from the Emperor of *Fez*, and *Morocco*, (who now reſides amongſt us) is reported to have advis'd his Attendants to ſee every thing, but admire nothing, left they ſhould ſeem thereby to diſparage their own Country, and ſhew themſelves ignorant of the great Rarities and Wonders of *Barbary*.   Poor

Poor contemptible *Berries*, fly hence to *Smyrna, Bantam,* or *Mexico,* then the Merchants would work through Storms and Tempests, through Fire and Water to purchase you, and at your Arrival here would proclaim your Vertues in all publick Assemblies; so true is that common saying, A Prophet is never valued in his own Country: The *English* Soyl is certainly influenced by some Pestilential Star, that blasts the credit of its Productions.

THE

The WAY of Making

M U M,

WITH SOME

REMARKS

UPON THAT

LIQUOR.

IN the firſt place, I will give ſome inſtructions how to make *Mum*, as it is Recorded in the Houſe of *Brunſwick*, and was ſent from thence to General *Monk*.

To make a Veſſel of 63 Gallons, the Water muſt be firſt boyl'd to the Conſumption of a third part, let it then be Brew'd according to Art with 7 Buſhels of Wheat-Malt, one Buſhel of Oat-Malt, and one Buſhel of Ground Beans, and when it is Tun'd, let not the Hogſhead be too much fill'd at firſt; when it begins to work, put to it of the inner Rind of the *Firr* three pounds, of the tops of *Firr*, and *Birch*, of each one pound, of *Carduus Benedictus* dried, three handfuls, Flowers of *Roſa Solis*, two handfuls,

handfuls, of *Burnet, Betony, Marjoram, Avens, Penny-royal,* Flowers of *Elder, Wild Thyme,* of each one handful and a half, Seeds of *Cardamum* bruised, three ounces, *Bayberries* bruised, one ounce, put the Seeds into the Veſſel; when the Liquor hath wrought a while with the Herbs, and after they are added, let the Liquor work over the Veſſel as little as may be, fill it up at laſt, and when it is ſtopped, put into the Hogſhead ten new laid Eggs, the Shells not cracked, or broken: ſtop all cloſe, and drink it at two years old, if carried by Water it is better.   Dr. *Ægidius Hoffman* added *Water Creſſes, Brooklime,* and *Wild Parſley,* of each ſix handfuls, with ſix handfuls of *Horſe Rhadiſh* raſped in every Hogſhead: it was obſerv'd that the *Horſe Rhadiſh* made the *Mum* drink more quick than that which had none.

By the compoſition of *Mum* we may gueſs at the qualities, and properties of it, you find great quantities of the Rind, and tops of *Firr* in it; therefore if the *Mum-*makers at *London* are ſo careful, and honeſt, as to prepare this Liquor after the *Brunſwick* faſhion, which is the genuine and original way; it cannot but be very powerful againſt the breeding of Stones, and againſt all Scorbutick Diſtempers.   When the *Suedes* carried on a Mollenbroc. *de Arthritide vag. Scorbut.* p.116. War againſt the *Muſcovites,* the Scurvy did ſo domineer amongſt them, that their Army did languiſh, and moulder away to nothing, till once encamping near a great number of *Firr Trees,* they began to boyl the tops of them in their Drink, which recover'd the Army even to a miracle; from whence the *Suedes* call the *Firr* the Scorbutick Tree at this very day.   Our moſt renowned Dr. *Walter Needham* has obſerv'd the great ſucceſs of theſe tops of *Firr* in the Scurvy, as Mr. *Ray* informs us; Ray *Catalog. Plantar.* which is no great wonder, if we conſider the Balſam, or *Turpentine,* (with which this Tree abounds) which proves ſo effectual in preſerving even dead Bodies themſelves

E   ſelves

selves from putrefaction, and corruption; if my memory does not deceive me, I have heard Mr. *Boyle*, (the ornament, and glory of our *English* Nation) affirm, that the Oyl of *Turpentine* preserves Bodies from Putrefaction much better than the Spirit of Wine. The *Firr* being a principal ingredient of this Liquor, is so Celebrated by some modern Writers, that it alone may be sufficient to advance the *Mum* trade amongst us. *Simon Pauli* (a Learned *Dane*) tells us the great exploits of the tops of this Tree in freeing a great man of *Germany* from an inveterate Scurvy; every Physician will inform you, how proper they are against the breeding of Gravel, and Stones: but then we must be so exact, as to pull these tops in their proper Season, when they abound most with *Turpentine*, and *Balsamick* parts, and then they may make the *Mum* a proper Liquor in *Gonorrhœa's*; besides the Eggs may improve its faculty that way: yet I will not conceal what I think the Learned Dr. *Merret* affirms in his Observations upon Wines, that those Liquors, into which the Shavings of *Firr* are put, may be apt to create pains in the Head; but still it is to be confess'd, that the *Firr* cannot but contribute much to the vigor and preservation of the Drink.

By the variety of its *Malt*, and by the ground *Beans*, we may conclude, that *Mum* is a very hearty and strengthning Liquor; some Drink it much, because it has no *Hops*, which they fancy do spoil our *English* Ales, and Beers, ushering in Infections; nay, Plagues amongst us. *Thomas Bartholine* exclaims so fiercely against *Hops*, that he advises us to mix any thing with our Drink rather than them: he recommends *Sage*, *Tamarisk*, tops of *Pine*, or *Firr*, instead of *Hops*, the daily use of which in our *English* Liquors is said to have been one cause, why the Stone is grown such a common Disease amongst us *Englishmen*: yet Captain *Graunt* in

his

Simon Pauli quadripart. Botan. p. 540.

Dr *Merrets* observations upon Wines.

Bartholine *de Medicinâ Danorum differtat.* 7.

Graunt's observations on the Bills of Mortality.

his curious Obfervations upon the Bills of Mortality, obferves, that fewer are afflicted with the Stone in this prefent Age, than there were in the Age before, though far more *Hops* have been us'd in this City of late than ever.

As for Eggs in the Compofition of *Mum* they may contribute much to prevent its growing fower, their fhells fweetning Vinegar, and deftroying Acids, for which reafon they may be proper in reftoring fome decay'd Liquors, if put whole into the Veffel. Dr. *Stubbs* in fome curious Obfervations made in his Voyage to *Jamaica*, affures us, that Eggs put whole into the Veffel will preferve many Drinks even to admiration in long Voyages: the Shells, and Whites will be devour'd and loft, but the Yolks left untouched. *[margin: Oldenburg's Philof. Tranf. act. N. 27.]*

Dr. *Willis* prefcribes *Mum* in feveral Chronical Diftempers, as Scurvies, Dropfies, and fome fort of Confumptions. The *Germans*, efpecially the Inhabitants of *Saxony*, have fo great a Veneration for this Liquor, that they fancy their Bodies can never decay, or pine away, as long as they are Lin'd, and Embalm'd with fo powerful a preferver; and indeed, if we confider the frame, and complexions of the *Germans* in general, they may appear to be living Mummies. But to conclude all in a few words, if this Drink call'd *Mum*, be exactly made according to the foregoing inftructions, it muft needs be a moft excellent alterative Medicine, the ingredients of it being very rare and choice fimples, there being fcarce any one Difeafe in Nature, againft which fome of them are not prevalent, as *Betony, Marjoram, Thyme*. In Difeafes of the Head; *Birch, Burnet, Water-Creffes, Brooklime, Horfe-Rhadifh* in the moft inveterate Scurvies, Gravels, Coughs, Confumptions, and all obftructions. *Avens*, and *Cardamom* Seeds for cold weak Stomachs. *Carduus Benedictus*, and *Elder* Flowers in inter- *[margin: Dr. Willis de Scorbuto. Pharmaceut. Rational. p. 2.]*

mittent Feavers. *Bayberries* and *Penny-Royal*, in Diftempers attributed to the Womb. But it is to be fear'd, that feveral of our *Londoners* are not fo honeft, and curious, as to prepare their *Mum* faithfully, and truly; if they do, they are fo happy as to furnifh, and ftock their Country with one of the moft ufeful Liquors under the Sun, it being fo proper, and effectual in feveral lingring Diftempers, where there is a depravation, and weaknefs of the Blood and Bowels.

There ftill remains behind a ftrong, and general objection, that may perhaps fall upon this little puny Pamphlet, and crufh it all to pieces, that is, the Hiftories are too fhort, and imperfect; to which I have only this to anfwer, *Ars longa, vita brevis,* a perfect Natural Hiftory of the leaft thing in World, cannot be the Work of one Man, or fcarce one Age, for it requires the Heads, Hands, Studies, and Obfervations of many, well Compar'd and Digefted together: therefore this is rather an Effay, or Topick, for men to reafon upon, when they meet together at Publick-Houfes, and to encourage them to follow the example of *Adam*, who being in the ftate of Innocence did contemplate of all the Creatures that were round about him in Paradife, but after the Fall, and the Building of a City, the Philofopher turn'd Politician.

Poſtſcript.

Liquors and Drinks are of ſuch general uſe, and
eſteem in all the habitable parts of the World,
that a word or two concerning them cannot
be improper, or unwelcom.

Firſt the Saps and Juices of Trees will afford many
pleaſant and uſeful Liquors.   The *Africans* and
Indians prepare their famous Palm Wine (which they
call *Sura*, or *Toddy*) out of the ſap of the wounded *Palm
Tree*, as we do our *Birch Wine* in *England* out of the
tears of the pierced *Birch Tree*, which is celebrated in *Helmont de Li-
thiaſi.*
the Stone and Scurvy. So the *Sycamore* and *Walnut* being *Ray's Catalog.*
wounded will weep out their Juices, which may be fer- *Plantar.*
mented into Liquors: In the *Molucca's* the Inhabitants *Vinetum Britannicum.*
extract a Wine out of a Tree called *Laudan.*

Fruits and Berries yield many noble and neceſſary
Liquors; every Nation abounds with various Drinks
by the diverſity of their Fruits and Vegetables. *England*
with *Sider, Perry, Cherry, Currant, Gooſeberry, Raſber-
ry, Mulberry, Blackberry,* and *Strawberry* Wine. *France,
Spain, Italy, Hungary* and *Germany,* produce great variety of Wines from the different ſpecies, and natures of
their Grapes and Soils.   In *Jamaica* and *Brazil* they *Vinetum Britannicum.*
make a very delicious Wine out of a Fruit called *Ananas,*
which is like a *Pine Apple,* not inferiour to *Malvaſia*
Wine.   The *Chineſes* make curious Drinks out of their
Fruits; ſo do the *Brazilians,* and Southern *Americans*; *Piſo Natur.
Hiſtor. India*
as from their *Coco, Acajou, Pacobi,* their *Unni,* or *Mur-*
tilla's.

POSTSCRIPT.

tilla's. We may note here, that all the Juices of Herbs, Fruits, Seeds, and Roots will work, and ferment themselves into intoxicating Liquors, out of which Spirits, and Brandies may be extracted, moſt Nations under the Sun has their drunken Liquors and Compounds; the *Turk* his *Maſlack,* the *Perſians* their *Bangue,* the *Indians* their *Fulo,* their *Rum,* their *Arak,* and *Punch.* The *Arabians, Turks, Chineſes, Tartars,* and other *Eaſtern* Countries do make inebriating Liquors out of their Corn, and Rice : ſome rather than not be Drunk will ſwallow *Opium, Dutroy,* and *Tobacco,* or ſome other intoxicating thing, ſo great an inclination has Mankind to be exalted. *Pliny* complains, that Drunkenneſs was the ſtudy of his time, and that the *Romans* and *Parthians* contended for the glory of exceſſive Wine Drinking. Hiſtorians tell us of one *Novellius Torquatus,* that went through all the honourable degrees of Dignity in *Rome,* wherein the greateſt Glory, and Honour he obtain'd, was for the Drinking, in the preſence of *Tiberius,* three Gallons of Wine at one Draught, before ever he drew his breath, and without being any ways concern'd. *Athenæus* ſays, that *Melanthius* wiſh'd his own Neck as long as a *Crane's,* that he might be the longer a taſting the pleaſure of Drinks ; yet what he reports of *Laſyrtes* is wonderful, that he never drank any thing, yet notwithſtanding Urin'd as others do. The ſame famous Author takes notice, that the great Drinkers us'd to eat *Coleworts* to prevent Drunkenneſs, neither are ſome men of our days much inferiour to thoſe celebrated Antients. The *Germans* commonly Drink whole Tankards, and Ell Glaſſes at a Draught, adoring him that Drinks fairly, and moſt, and hating him that will not pledg them. The *Dutch* Men will ſalute their Gueſts with a Pail, and a Diſh, making Hogſheads of their Bellies. The *Polander* thinks him the braveſt fellow, that Drinks moſt Healths, and car-

ries

POSTSCRIPT.

ries his Liquor beſt, being of opinion, that there is as much Valour in Drinking, as Fighting. The *Ruſſians, Suedes,* and *Danes,* have ſo naturaliz'd *Brandy, Aqua Vitæ, Beer, Mum,*&c. that they uſually Drink our *Engliſh* Men to Death, ſo that the moſt ingenious Author of the *Vinetum Britannicum* concludes, that temperance (relatively ſpeaking) is the Cardinal Vertue of the *Engliſh.*

It is very wonderful what Mr. *Ligon,* and other *American* Travelers relates of the *Caſſava Root,* how out of it, the *Americans* do generally make their Bread, and common Drink, called *Parranow;* yet that Root is known to be a great Poyſon if taken raw: their Drink call'd *Mobby* is made of *Potatoe's.* But we will conclude all with *Virgil,* who ſpeaking of the many Liquors in his time, ſays, *Sed neque quam multæ ſpecies, nec non quæ ſunt eſt Numerus.*

Ligon's Hiſtory of Barbados.

F I N I S.

A Help to *English* History, containing a Succeffion of all the Kings of *England*, the *English Saxons*, and the *Britains*; the Kings and Princes of *Wales*, the Kings and Lords of *Man*, the Ifle of *Wight*: As alfo containe the Dukes, Marqueffes, Earls and Bifhops thereof: With the Defcription of the places from whence they had their Titles: Together with the Names, and Ranks of the Vifcounts, Barons and Baronets of *England*. By *P. Heylyn*, D. D.

Monumenta Weftmonafterienfia: Or an Hiftorical Account of the Original, Increafe, and Prefent State of St. *Peter*'s, or the Abby Church of *Weftminfter*. With all the Epitaphs, Infcriptions, Coats of Arms, and Atchievments of Honour belonging to the Tombs and Grave-ftones: Together with the Monuments themfelves faithfully defcribed and fet forth. By *H. K.* of the Inner Temple, Gent.

www.ingramcontent.com/pod-product-compliance
Lightning Source LLC
Chambersburg PA
CBHW021447090426
42739CB00009B/1679